DANGEROUS SNAKES

RATTLESNAKES

Tracy Nelson Maurer

TABLE OF CONTENTS

A Crabtree Seedlings Book

CRABTREE
Publishing Company
www.crabtreebooks.com

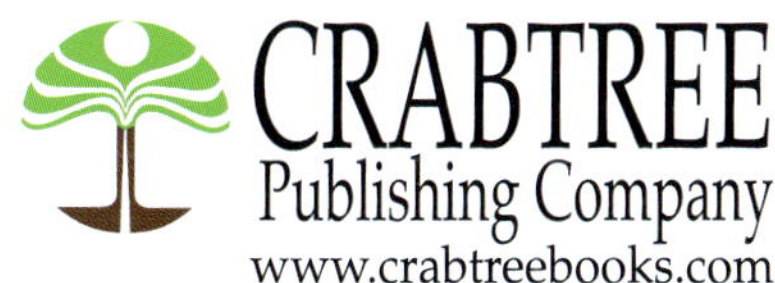

School-to-Home Support for Caregivers and Teachers

This book helps children grow by letting them practice reading. Here are a few guiding questions to help the reader with building his or her comprehension skills. Possible answers appear here in red.

Before Reading:

- What do I think this book is about?
  - *I think this book is about rattlesnakes.*
  - *I think this book is about why rattlesnakes make rattling sounds.*
- What do I want to learn about this topic?
  - *I want to learn how to stay away from rattlesnakes.*
  - *I want to learn where rattlesnakes live.*

During Reading:

- I wonder why…
  - *I wonder why rattlesnakes don't like cold weather.*
  - *I wonder why some people eat rattlesnakes.*
- What have I learned so far?
  - *I have learned that rattlesnakes must shed their skin to grow.*
  - *I have learned that they add a ring to their rattles each time they shed their skin.*

After Reading:

- What details did I learn about this topic?
  - *I learned that rattlesnakes eat mice, birds, and other snakes.*
  - *I learned that rattlesnakes don't like cold air.*
- Read the book again and look for the vocabulary words.
  - *I see the word **rattler** on page 5 and the word **scales** on page 4. The other glossary words are on pages 22 and 23.*

RATTLESNAKES

Ch-ch-ch-ch-ch-ch!

A rattlesnake shakes its tail to say, “Move away!”

Hard **scales** form the rattle.

A **rattler** can shake its tail more than 50 times per second.

Rattlesnakes must shed their skin to grow.

Rattlesnakes add a ring to their rattles each time they shed their skin.

Rattlesnakes may look tan, brown, or gray.

Rattlesnakes wait for **prey**.

Pits on their snouts help them track their prey's body heat.

Rattlers eat mice, birds, and other snakes.

Sharp **fangs** squirt deadly **venom**.

These **reptiles** stay as warm or as cool as the air around them.

Cold air slows rattlesnakes. They often warm their bodies in the morning sunshine.

Most rattlesnakes live in rocky deserts or forests.

Some people eat rattlesnakes. Would you?

Glossary

fangs (FANGZ): Fangs are sharp, long teeth.

prey (PRAY): Prey is an animal that is hunted by another animal for food.

rattler (RAT-lur): A rattler is a nickname for a rattlesnake.

reptiles (REP-tilez): Reptiles are cold-blooded, scaly animals that breathe air.

scales (SKAYLZ): Scales are small pieces of hard skin that cover the bodies of reptiles and fish.

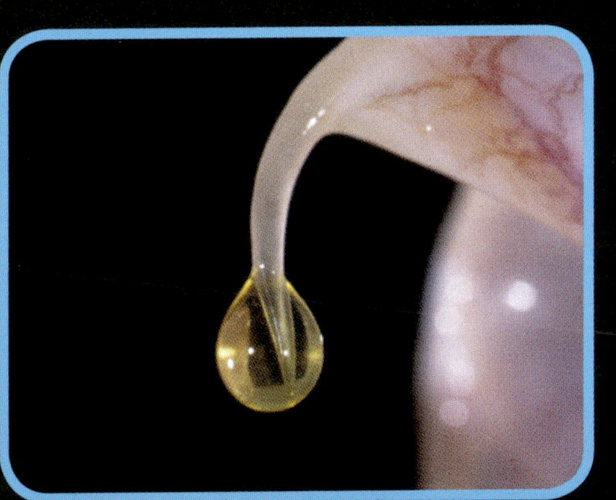

venom (VEN-uhm): Venom is a poison passed through a bite or sting.

Index

About the Author

Tracy Nelson Maurer

Tracy Nelson Maurer has written more than 100 books for young readers. She lives in Minnesota where it's too cold for most dangerous snakes.

Websites

https://kids.nationalgeographic.com/animals/reptiles/rattlesnake/
https://easyscienceforkids.com/all-about-rattlesnakes/

Written by: Tracy Nelson Maurer
Designed by: Jennifer Dydyk
Edited by: Kelli Hicks

Photographs:
mask for snakeskin graphic on cover and pages © shutterstock.com/Merydolla. yellow triangle with snake graphic © Top Vector Studio/Shutterstock; Cover photo: © Steve Byland/Shutterstock; page 3 © Audrey Snider-Bell/Shutterstock; page 5 © Creeping Things/Shutterstock; page 7 © Chaikom/Shutterstock; page 9 © Susan Schmitz/Shutterstock; page 11 © Ali Iyoob Photography/Shutterstock; page 13 © Breck P. Kent/Shutterstock; page 15 © Audrey Snider-Bell/Shutterstock; page 17 © Edgar Figueiredo/Shutterstock; page 19 © Brian Magnier/Shutterstock; page 21 © Brian Magnier/Shutterstock; page 23 (bottom photo) © Joe McDonald/Shutterstock

Library and Archives Canada Cataloguing in Publication

Title: Rattlesnakes / Tracy Nelson Maurer.
Names: Maurer, Tracy Nelson, 1965- author.
Description: Series statement: Dangerous snakes | "A Crabtree seedlings book". | Includes index.
Identifiers: Canadiana (print) 20210202181 | Canadiana (ebook) 2021020219X | ISBN 9781427162373 (hardcover) | ISBN 9781427159274 (softcover) | ISBN 9781427159281 (HTML) | ISBN 9781427159298 (EPUB) | ISBN 9781427159304 (read-along ebook)
Subjects: LCSH: Rattlesnakes—Juvenile literature.
Classification: LCC QL666.O69 M39 2022 | DDC j597.96/38—dc23

Library of Congress Cataloging-in-Publication Data

Names: Maurer, Tracy Nelson, 1965- author.
Title: Rattlesnakes / Tracy Nelson Maurer.
Description: New York : Crabtree Publishing, [2022] | Series: Dangerous snakes- a Crabtree seedlings book | Includes index.
Identifiers: LCCN 2021018833 (print) | LCCN 2021018834 (ebook) | ISBN 9781427162373 (hardcover) | ISBN 9781427159274 (paperback) | ISBN 9781427159281 (ebook) | ISBN 9781427159298 (epub) | ISBN 9781427159304
Subjects: LCSH: Rattlesnakes--Juvenile literature.
Classification: LCC QL666.O69 M38 2022 (print) | LCC QL666.O69 (ebook) | DDC 597.96/38--dc23
LC record available at https://lccn.loc.gov/2021018833
LC ebook record available at https://lccn.loc.gov/2021018834

Crabtree Publishing Company
www.crabtreebooks.com 1-800-387-7650

Printed in the U.S.A./062021/CG20210401

Published in the United States
Crabtree Publishing
347 Fifth Avenue, Suite 1402-145
New York, NY, 10016

Published in Canada
Crabtree Publishing
616 Welland Ave.
St. Catharines, Ontario L2M 5V6